EDITOR: MARTIN WINDROW

OSPREY MILITARY · **MEN-AT-ARMS SERIES** · 253

WELLINGTON'S HIGHLANDERS

Text by
STUART REID
Colour plates by
BRYAN FOSTEN

D1471423

First published in Great Britain in 1992
by Osprey, an imprint of Reed Consumer Books Limited
Michelin House, 81 Fulham Road
London SW3 6RB
and Auckland, Melbourne, Singapore and Toronto

ISBN 1 85532 256 0

Printed through Bookbuilders Ltd, Hong Kong

Artist's Note
Readers may care to note that the original paintings
from which the colour plates in this book were
prepared are available for private sale. All
reproduction copyright whatsoever is retained by the
publisher. All enquiries should be addressed to:

Bryan Fosten
5 Ross Close
Nyetimber
Nr. Bognor Regis
Sussex PO21 3JW

The publishers regret that they can enter into no
correspondence upon this matter.

Dedication
To the memory of Lieutenant John Urquhart, 71st
Highlanders, died Negapatam, Southern India, 24
April 1794.

Acknowledgements
I would like to thank the staff of the Scottish United
Services Museum in Edinburgh Castle for their
assistance in preparing this study. The SUSM
collections deserve to be far better known. I would also
like to thank Lt.Col. A. A. Fairrie, of RHQ the Queen's
Own Highlanders, for his advice on the status of
pipers, although the conclusions presented here are my
own.

Bibliography
A number of manuscript and printed sources were
consulted in the preparation of this study, most
notably: David Stewart of Garth's *Sketches of the
Highlanders of Scotland* (Edinburgh 1822); Diana M.
Henderson's *The Highland Soldier 1820-1920*
(Edinburgh 1989) provides a rather more mature
assessment of the character of highland regiments than
John Prebble's polemical *Mutiny* (London 1975);
J. M. Bulloch's *Territorial Soldiering in the North-East
of Scotland 1759-1815* (Aberdeen 1914) is quite
invaluable.

RAISING A REGIMENT

On the outbreak of war with France on 1 February 1793, six of the British Army's 77 regiments of infantry (excluding the Footguards) bore the designation 'Highland'; and within two years this number had risen to a temporary high of 15, besides a number of regiments of Highland Fencibles. Ironically some of the latter, raised originally as home defence battalions, saw more action, in Ireland, than some of the short-lived regular units.

As a first step to raising a regiment a prospective colonel had to obtain 'letters of service', authorising him to nominate officers and beat up for recruits. Although there were variations, those for the 100th (Gordon) Highlanders, granted to the Duke of Gordon on 10 February 1794, may be taken as typical:

'My Lord,

'I am commanded to acquaint you that His Majesty approves of your Grace's offer of raising a Regiment of Foot to be completed within three months upon the following conditions:-

'The Corps to consist of one company of Grenadiers, one of Light Infantry, and eight Battalion Companies.

'The Grenadier Company is to consist of one captain, three lieutenants, four sergeants, five corporals, two drummers, two fifers, and 95 privates.

'The Light Infantry of one captain, three lieutenants, four sergeants, five corporals, two drummers, and 95 privates.

'And each Battalion Company of one captain, two lieutenants, one ensign, four sergeants, five corporals, two drummers, and 95 privates;

'Together with the usual staff officers, and with a sergeant major, and quartermaster sergeant,

Dayes watercolour, c.1791, depicting a soldier of the 42nd. The subject's rank is unclear: the elaborate purse, pistol, broadsword and dirk all suggest a sergeant, but the red line in the lace loops on the coat and the absence of epaulettes indicate a private or corporal. It is possible that the soldier has been 'got up' for a quarter guard. Also of interest are the knitted stockings, identified as such by the turned-over tops and absence of a black edging to the check. (Scottish United Services Museum)

exclusive of the sergeants above specified. The captain lieutenant is [as usual] included in the number of lieutenants above mentioned.

'The corps is to have three field officers, each with a company; their respective ranks to be determined by the rank of the officer whom your Grace will recommend for the command thereof. If the person so recommended for the command is not at present in the army, he will be allowed temporary rank during the continuance of the regiment on the establishment but will not be entitled to half pay on its reduction.

'His Majesty leaves to your Grace the nomination of all the officers, being such as are well affected to his Majesty, and most likely by their interest and connections to assist in raising the corps without delay; who, if they meet with his Royal approbation, may be assured they shall have commissions as soon as the regiment is completed.

'The officers, if taken from the half pay, are to serve in their present ranks; if full pay, with one step of promotion. The gentlemen named for ensigncies are not to be under sixteen years of age. The quartermaster is not to be proposed for any other commission.

'In case the corps shall be reduced after it has been established, the officers will be entitled to half pay.

'The pay of the officers is to commence from the dates of their commissions: and that of the non-commissioned officers and private men from the dates of their attestations.

'Levy money will be allowed to Your Grace in aid of this levy at the rate of five guineas per man for 1064 men.

'The recruits are to be engaged without limitation as to the period or place of their service.

'None are to be enlisted under five feet four inches, nor under 18 years or above 35. Growing lads from 16 to 18, at five feet three inches will not be rejected.

'The non-commissioned officers and privates are to be inspected by a general officer, who will reject all such as are unfit for service, or not enlisted in conformity with the terms of this Letter.

'His Majesty consents that on a reduction the Regiment shall, if it be desired, be disbanded in that part of the country where it was raised.

'In the execution of this service, I take leave to assure Your Grace of every assistance which my office can afford.

Sir George Yonge'

The Officers

Having thus secured his letters of service, a colonel had next to find the officers to whom he would sub-

Government or 'Black Watch' tartan. This sett has a blue base overlaid with a broad green check bordered by mixed black/blue checks and divided by a black overstripe. This black overstripe was replaced in some regiments by a red, white, buff or yellow one. The blue base is additionally overlaid with fine black lines and in the case of the McKenzie sett, used by the 71st and 78th, an additional red overstripe is laid through alternate blue squares. (Author)

contract the 'horrible drudgery' of recruiting. The captains and field officers were almost invariably promoted from other regiments, or appointed from the Half Pay list, though in highland regiments efforts were always made to attract men with local roots. Otherwise it was a matter of calling upon an extensive network of neighbours and tenants, who could assist in recruiting and might occasionally even take up commissions themselves. They could also recommend deserving young members of their own or acquaintances' families fo fill the junior vacancies. All of the original officers commissioned into the 100th (Gordon) Highlanders were Scots, and most of them came from the areas north and west of Aberdeen, where the regiment was primarily recruited. In marked contrast we may consider, for instance, the 106th (Norwich) Regiment, typical of the English units raised in 1794, which had remarkably few Norfolk officers. A surprising number of them – including an ancestor of the author – were Scots, with no local ties or influence whatever.

Highland regiments were frequently raised by chieftains, but they were not by any means *clan* regiments. Sir James Grant's 97th Highlanders had some 14 officers (just under half) named Grant on its original establishment, but this appears to have been exceptional, and most of them were in any case subalterns. More typical were the Gordons: only five officers, including the colonel, bore that surname, and this seems to have been the usual proportion. Both battalions of the 78th, raised in 1793 and 1794, had five McKenzies apiece; and only four of the 79th, raised in 1793, were named Cameron.

Enlistment

In order to qualify for the commissions for which they were recommended the officers had in turn to enlist the required number of men. Naturally they sought to do so close to home, where they were 'weel kenn'd' and where they could command some influence. While an officer might reasonably expect to find a few recruits amongst the sons of his tenants (if he possessed any), most recruits were to be found at local markets, or were encouraged to enlist by the officers' friends and relations.

The 93rd, it is true, were substantially raised on the Sutherland estates by a form of conscription, but this was evidently exceptional and in the long term

Lieutenant Aeneas Sutherland, 2/78th Highlanders, who was commissioned 27 March 1794. Note the buff square at the intersection of the red checks on the bonnet dicing, possibly a 2nd Bn. distinction; a bugle horn badge on the epaulettes; and, oddly, silver lace – regiments with buff facings were normally unlaced. (SUSM)

unsustainable. More typical was the sort of assistance given to officers beating up for the Gordons. On 28 April 1794 the *Aberdeen Journal* reported that the Duke's tenants in Kirkmichael and Strathdon had met and agreed to find an additional bounty of three guineas for any man from the area enlisting in the regiment. Similarly the Earl of Aberdeen and Mr. Skene of Skene were reported in the following month to be encouraging men from their estates to enlist with the rival 109th Highlanders, and the former also provided large quantities of the whiskey punch which seems to have been so essential for successful recruiting. (Captain Finlason, the Gordons' agent in Aberdeen, complained on 24 June that his house was like an alehouse or gin shop.)

In the north of Scotland 'going for a soldier' had none of the stigma which attached to it in England at

Main Gate, Fort George, Ardersier. This large fort and barrack complex, completed in 1769, could accommodate two battalions; and at one time or another most highland regiments were inspected or garrisoned there. (Author)

this time. The highlands were badly over-populated and enlistment in the army was widely regarded as an entirely respectable alternative to emigration, particularly since the unusually high bounties on offer meant that some measure of provision could be made for those relatives left behind. The bounties were high not because it was hard to persuade men to enlist in the first place, but rather because they were necessary to encourage a man to enlist in, say, the 100th rather than the 109th. Consequently recruiters were constantly frustrated by the fact that the men they sought were inclined at first to hang back, not through any unwillingness to enlist or because, as is sometimes suggested, they were turning against a vestigial clan system; but rather because they not unnaturally looked around for the best offer available before committing themselves. The point is cheer-

fully illustrated by a surviving recruiting banner for the 2/78th which bears little more than the words *'SIXTEEN GUINEAS BOUNTY'*.

Notwithstanding the high bounties it was always necessary to enlist at least some men from further afield to make up the required numbers, though the proportions varied from regiment to regiment according to circumstances. Generally speaking this involved the recruiting of lowland Scots, rather than the Irishmen who were the usual standby in English units. Irishmen in fact only accounted for about 6% of the strength of highland regiments, in marked contrast to the 30% or more frequently found in English ones.

Of the 894 men originally recruited into the Gordons in 1794 whose origins are known, only 390 were highlanders; another 407 were lowland Scots – reflecting the high proportion of lowland officers – while 51 were Irish, 11 were English or Welsh, and there was a solitary German musician, Carl Augustus Rochling, from Hesse Cassell. The 78th, almost entirely staffed with highland officers, did rather better, and when its two battalions were consolidated in June 1796 they mustered 970 highlanders, 129 lowland Scots, with 14 English and Irish. The 78th appear to have had a prejudice against Irish recruits, and in January 1820 had only two Irishmen out of 640 rank and file.

Problems did arise, on the other hand, when regiments had to be virtually rebuilt after service overseas. In 1799 the 79th, raised afresh after drafting in Martinique, had only 268 Scots in the ranks, not all of them by any means highlanders, and no fewer than 273 Englishmen, 54 Irishmen and seven assorted foreigners. Under threat of losing their cherished *Highland* distinction the balance was rapidly re-dressed; but it is significant that most of the regiments which did lose their kilts in 1809 were in similar case, having spent upwards of 20 years in India.

The character of the regiments

Highland regiments were therefore much more homogenous than comparable English ones, such as the 106th, and local initiatives such as those reported in the *Aberdeen Journal* clearly encouraged the enlistment of men of good character, from particular districts. Their behaviour was consequently rather

better than usual. The Light Company of the 93rd was reputed not to have had a single man punished in 20 years. This is not to say that crime was unknown in highland regiments – a soldier who doesn't plunder is a contradiction in terms – but by and large memoirs refer to individuals misbehaving rather than to any endemic problem. There were, of course, lapses. Over the winter of 1810–1811, for example, the 71st and 92nd were billeted in the Convent of Alcantaria, and predictably soon tore down the screens and other woodwork for their fires. The 71st apparently took to this with some glee, and an officer of the Gordons was astonished one night to find them tossing images into the fire with cheerful cries of 'No Popery!'

Strathspey Fencibles drum; varnished wood shell with a light cream panel bearing a crown and a red shield with gold lettering and foliate decoration. Although none too clear on the print the shield has a large 'GR' cypher over 'STRATHSPEY FENCIBLES'. The hoops are red. The other equipment in the photo belonged to the 97th Highlanders; the knapsack is yellow ochre with a green ring bearing the title 'INVERNESS-SHIRE REGT.' in white letters and black 97 in the centre (see Plate D1). Note the flat section of the early pattern cartouche box. (SUSM)

OR's breastplate, Strathspey Fencibles. The officers' pattern was similar but of rather better quality. (Author)

Highlanders have also been represented, quite unjustly, as being prone to mutiny in protest against cynical exploitation by unscrupulous officers. Incidents, occasionally violent, certainly occurred in some of the Fencible regiments in the early 1790s, and there had earlier been some trouble during the American War; but it is important that such outbreaks should not be considered in isolation. The commonest cause of mutiny in the 18th century British Army was drafting or some similar real or imagined breach of a regiment's terms of service; and in such circumstances English and Irish regiments mutinied equally readily and frequently more violently. A number of the English and Irish regiments involved in the great drafting of August 1795 mutinied, but none of the five highland regiments also ordered to be drafted at that time misbehaved.

Ultimately, regiments are judged by their behaviour in battle; and highlanders, whether recruited in Lochaber or the lowlands of Aberdeenshire, have always had a reputation as 'stormers', exemplified by the impetuous charge of the Gordons at Waterloo, intermingled with the Scots Greys. This reputation probably resulted at least in part from an unusually close bonding between officers and men, and an assumption that highlanders were natural soldiers, possessed of an impetuous spirit and temperamentally more inclined to use the bayonet. Whether justified or not this image was in large measure self-perpetuating, since exactness in drill seems sometimes to have been neglected as being unnecessary, as Sgt. Anton relates:

'First, the ruggedness of the mountains prevented precision of movements; secondly, the weather had become so unfavourable that every fair day was dedicated to some other necessary purpose about the camp, and instead of acquiring practical knowledge himself [Colonel Macara of the 42nd], even his regiment was losing part of that which it had perhaps previously possessed; thirdly, draughts of undisciplined recruits were occasionally joining and mixing in the ranks, and being unaccustomed to field movements, occasioned a sort of awkwardness in the performance of them. Even after our return from the continent, when the regiment was quartered in Ireland, many obstacles started up unfavourable to field practice, namely old soldiers and limited service men being discharged, the second battalion joining, the principal part of which were recruits, and men who had been years in French prisons; the detached state of the regiment, after all these had been squad drilled, left but few soldiers at headquarters to enable the commanding officer to practise with. In this manner we continued until the battle of Waterloo. ... We had the name of a *crack* corps, but certainly it was not then in that state of discipline which it could justly boast of a few years afterwards.'

When the first highland regiments were raised in the mid-18th century they were regarded as a species of light infantry, and frequent comparison was made with Pandours and other irregulars; but by the 1790s they were simply used as line infantry. Nevertheless the tradition of self-sufficiency died hard. Battalion companies were sometimes ordered out with the Light Company, to thicken the skirmish line. The practice was not unknown in English regiments; but highlanders also maintained the system of selecting individuals from the Battalion companies who were good shots, to serve as additional skirmishers or *Flankers*. This was last done by English regiments at Maida and was positively forbidden by Sir John Moore shortly afterwards, since he considered that it unduly weakened the firing line. The 79th, at least, appear to have ignored him, and according to Lt. Alexander Forbes of No. 1 Coy. the Light Companies of Kempt's Brigade were reinforced by 'the 8th Company and marksmen of the 79th Regiment' at Quatre Bras.

In the light of this it is perhaps not surprising that four out of the 13 regiments providing personnel for the Experimental Rifle Corps in 1800 were 'Highland': 71st, 72nd, 79th and 92nd. In 1805 the process was carried a stage further when the newly raised 2/78th was selected by Sir John Moore for training as a Light Infantry regiment, and actually spent some months at Shorncliffe. Before their training was complete, however, they had to be posted with the 1/42nd to Gibraltar and thereafter served as line infantry, the light infantry role eventually being taken up by the 71st.

HIGHLAND UNIFORM

The **kilt**, the most distinctive part of highland dress, comprises a single strip of tartan material with a central pleated section. It can be tailored by varying the number of pleats, although modern practice tends towards a standard number, varying from regiment to regiment. The kilt is worn with the pleated section at the rear while the two flat ends are folded one over the other, the left hand one uppermost, to form a double apron at the front.

Plaiding is traditionally woven in a 27 inch width, and since the kilt is so arranged that the bottom edge hangs an inch off the floor when kneeling, at this period the top edge usually came rather high up the body, providing excellent protection for the abdomen and kidneys in cold weather.

Surviving 18th century documentation relating to the provision of material for kilts indicates that $3\frac{1}{2}$ yards was considered sufficient for a soldier and 4 yards for a sergeant. In contrast a modern kilt normally requires 7 or 8 yards of material, though this is largely swallowed up in the deep folds of knife-pleating. Two surviving ORs kilts worn by the Gordons in the 1790s, however, are box-pleated. Since box pleats are no more than squared-off gathers very much less material is required, and the $3\frac{1}{2}$ yards quoted are entirely adequate. No lining was provided.

Officers, of course, were not so restricted as to the quantity of material used, and knife-pleating was certainly in use by the 1820s if not rather earlier. While the two methods of pleating are quite dissimilar, the difference cannot be discerned at any great distance.

In place of the modern arrangement of three straps and buckles fastening the kilt, reliance at this

Regimental colour, Strathspey Fencibles: green with gold wreath and thistle, the title in gold/black lettering on a red scroll. The presentation of these colours to the regiment at Linlithgow on 21 March 1794 was marred by a mutiny when some of the soldiers wrongly assumed that the ceremony presaged their being sent overseas. (Author)

time was placed upon pins (indeed, the Black Watch continued to fasten their kilts with three pins down either side of the apron until about 1914). This method did, however, have its drawbacks. There are occasional references to highlanders losing their kilts while forcing their way through hedges, abattis or anything else which might catch them. Sgt. Anton of the 42nd recounts an amusing instance of this which also answers the perennial question:

'The colonel ... strode hastily forward to enforce obedience; Doury was the first to observe him, fled past his companion, dropping the sticks at his feet and escaped. Not so Henderson; he fell over the bundle dropped at his feet with his face pressed against the soft miry field; the colonel overtook him as he recovered, seized him by the kilt, the pins of which yielded to the tug, and left his naked posteriors to some merited chastisement.'

The Gordons kilt in the SUSM, on the other hand, is fastened with two small flat brass buttons set into the very top at either side of the apron. There is some evidence of damage to the material caused by the prolonged use of pins, and the buttons seem to have been an expedient intended to avoid further damage.

The kilt had originally been used as a fatigue garment, but for all practical purposes it replaced the old belted plaid by the 1790s, though the appearance of the latter was still preserved on the parade ground by the **feile** or 'fly' plaid (little plaid). Comprising that part of the soldier's allowance of tartan material (6 yards every two years) left over after his kilt had been made, it was fastened from behind by one corner to the left shoulder strap button and otherwise left hanging free.

Officers appear to have continued to wear the full belted plaid, albeit with pre-sewn pleats, for formal parades at least until the early 1800s. At other times, particularly when wearing breeches, they too wore a slightly longer version of the fly plaid known as a 'highland scarf'. This was fastened behind to the left shoulder strap button as before, but instead of hanging free it was passed under the right arm and

back up across the body before being thrown over the left shoulder. When worn in this manner the plaid almost completely obscured and sometimes replaced the crimson silk net sash which was also worn over the left shoulder, rather than tied around the waist as in the rest of the army.

Breeches were worn surprisingly frequently, often in undress, though field officers, including the adjutant, wore them as a matter of course since their duties required them to be mounted. During the Peninsular War company officers were also encouraged to ride upon the march, and since the kilt is ill-suited to riding this effectively led to its falling completely into disuse on active service. Blue or grey overalls became the preferred wear, frequently with red stripes down the outside leg, but web pantaloons and half boots are also shown or recorded as being worn in the 71st and 92nd.

In the 1790s both officers and men of some regiments wore tartan pantaloons, or **trews** as Sir John Sinclair insisted on calling them. Strictly speaking, being footless they were nothing of the sort, but the name has stuck. During the Peninsular War they re-appeared, sometimes at a regimental level, but more usually informally. In 1814 the ever observant Sgt. Anton noted that the 42nd, 'the only corps in the brigade [1/42nd, 1/79th & 91st] that wore the kilt was beginning to lose it by degrees; men falling sick and left in the rear frequently got the kilt made into trowsers, and on joining the regiment again no plaid could be furnished to supply the loss; thus a great want of uniformity prevailed. . . .'

The practice of having kilts made into trousers while in rear areas seems to have been a long standing one, although grey trousers were officially allowed as a comfortable fatigue dress until about 1825, when officially replaced by trews. In late October or early November 1813 'T.S.' of the 71st noted that the 92nd had grey trousers served out to them since the weather was so cold. The 79th may also have received grey trousers at this time, since Anton is positive that they were no longer wearing the kilt in 1814. Tartan trousers appear, however, for obvious reasons to have

Broadsword hilt, Breadalbane Fencibles. This unusual style was also employed, with a minor alteration to the cartouche bearing the regimental badge, for the 116th Highlanders. (SUSM)

been more popular, and most of the 1/93rd wore tartan rather than grey trousers at New Orleans in 1815.

With the leg covered only to the knee by the kilt, **hose** formed an important part of highland uniform. Traditionally they were white with a simple pink/red check set diagonally (edged in black for the 42nd, 73rd and 78th), although the Dumbarton Fencibles had a grey/black check, matching their black facings. Originally they were fashioned out of woven cloth; but correspondence from the suppliers, William Wilson & Sons of Bannockburn, reveals that knitted hose were in use by the early 1800s, and some illustrations show them worn much earlier. Knitted hose can be easily identified in portraits and other contemporary illustrations by all-pink feet, a turned-over top, and, in the case of the 42nd, an absence of the black edging to the check.

The introduction of canvas half-gaiters for marching in 1800 also led to the unofficial adoption of **moggans** or footless hose, described by Sgt. Anton:

'. . . these are hose from which the feet have been cut, the spats cover this deficiency, and the legs appear as if the under parts were complete. This was an excellent method for hardening the feet so as to accustom them to bad shoes and bad roads . . . very few of the men wore any other kind, although they had complete ones in their possession.'

The hose, of whatever kind, were tied below the knee with scarlet worsted garters, generally with an elaborate knot, although some illustrations show rosettes. If knitted hose were worn the turned-over top covered the garter.

In the early 1790s all highlanders wore cropped light infantry style **jackets** with ten loops of regimental lace on each lapel, vertical false pocket flaps on the skirts, and only the front corners of the skirts turned back. The lapels served to display the regimental facing colour and, except by means of hooks and eyes immediately below the throat, the jacket could not be closed over the front of the body; instead it was cut away sharply to expose the white waistcoat. Although elegant this style was rather impractical, and after a brief experiment with closing the jacket down to the waist the army adopted a short single-breasted jacket in 1797, similar to that already used as slop clothing for recruits or for troops serving in the Indies. This new jacket was not an unmixed blessing for highlanders, as Stewart of Garth comments:

OR's breastplate, Reay Fencibles. The construction is interesting in that the badge and lettering are moulded in relief rather than incised. (Author)

'This was an improvement in the English uniform, as it gave additional warmth to the back and bowels; but when it was adopted by Highland corps, the nature of the garb was overlooked. The numerous plaits and folds of the belted plaid and little kilt form so thick a covering, that when the coat is added, the warmth is so great, that on a march it debilitates those parts of the body, whereas the former cut of the jacket, with the skirts thrown back, and the breast open, left them uncovered. . . .'

The new highland jacket, therefore, while otherwise conforming to the line pattern, was made shorter and had only eight instead of ten buttons at the front. There should also have been only three buttons on the false pockets, but illustrations and surviving examples reveal four buttons as on the old jacket. The white waistcoat could still be worn underneath the jacket for additional warmth in cold weather, but was now given sleeves and a plain white stand-up collar, effectively converting it into an undress 'stable jacket'.

Officers' jackets were originally similar to those worn by the rank and file, though obviously of superior quality and bearing gold or silver bullion epaulettes on each shoulder. When the single-

Regimental colour, 97th Highlanders: green with red disc, edged gold, black lettering and natural embroidered wreath. This is a rather unusual design for highland regiments and quite unlike that of the Strathspey Fencibles, also raised by Sir James Grant. (Author)

breasted jacket was introduced officers adopted instead a double-breasted one which could be properly buttoned over, or worn open to display the lapels and sometimes the waistcoat as well.

Officers and men wore Kilmarnock **bonnets**, so named from their principal place of manufacture. These were not made up from pieces of cloth like an English forage cap but were instead heavily knitted, felted and padded, drum-shaped blue woollen caps about six inches high. Most were produced by James MacLean of Kilmarnock and supplied as usual through William Wilson & Sons of Bannockburn, the principal outfitters of highland regiments. In 1813 officers' bonnets cost 36 shillings a dozen and sergeants' 28s., while ORs' cost only 21s. (in today's currency, £1.80, £1.40 and £1.05).

Bonnets were sometimes plain blue, but normally had a three-inch-deep diced band featuring a red check on a white ground. The origin of this band has been the subject of considerable discussion but the likeliest explanation is that it was simply a decorative addition, mirroring the diced hose. It normally had green squares at the intersection of the red checks, although variations are recorded, most notably the

white squares of the 'Sutherland' dicing. The rear of this band had a deep 'V'-shaped slit and the bottom rim of the bonnet, below the band, was doubled over to enclose a tightening tape or ribbon, tied at the rear with a large bow. When the ribbon was untied the bonnet could, according to Stewart of Garth, be pulled down like a nightcap, and this was sufficiently comfortable and popular to attract complaints in standing orders. In 1796, for example, officers of the Gordons were rather sniffily instructed to ensure that their men's bonnets had black ribbons at the rear so that they could be worn 'properly, and not down on their heads like a nightcap'. Nevertheless photographs taken of members of the regiment in 1845 still show the bonnet pulled down.

In the centre of the crown, which tended to sag, there was a tufted woollen ball or 'Tourie': white for grenadiers, red for battalion companies and green for light infantry. On the left side was a circular cockade about three inches in diameter. These cockades were normally black but the grenadiers of the 42nd had red ones and the light infantry green, edged red, while the Gordons' light company had red cockades. Initially they were secured in the centre by a regimental button, but after 1801 other devices such as sphinx badges, grenades and bugle horns also appeared.

For most duties Kilmarnock bonnets were supposed to be 'mounted' with black ostrich feathers fastened in the left side by the cockade and falling over the crown. A regimental order of the 92nd dated 7 March 1799 specified one large and six small feathers. This was not invariably the case, however. In 1794 Col. Lewis McKenzie of the 2/78th issued imitation feathers in worsted. He reasoned that they could thus be made to look uniform in size and shape: 'Every one knows who is acquainted with the 42nd Regiment that their bonnets are the least uniformly dressed, and the worst part of their clothing and appointments.' It may be no coincidence that an inspection report on the 42nd, of 14 May 1790, had complained that the bonnets were 'entirely disfigured, they are so covered with lofty feathers that they appear like high Grenadier caps of black bearskin, and are by that means expensive to the men'. Dayes' watercolour painted in the following year shows relatively few feathers, perhaps in response to the previous year's report, but their arrangement certainly appears less than elegant.

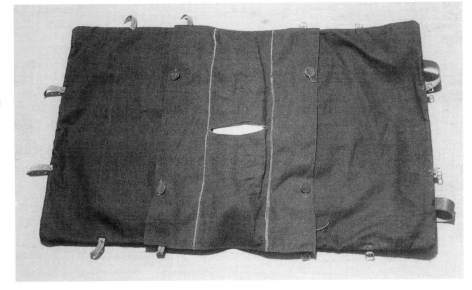

Internal and external views of the so-called 'envelope' knapsack, as used by the Strathspey Fencibles c. 1793. Essentially it comprises two large pockets with a smaller one, for blackballs and other dirty items, accessed from the central slit. Internal stitching is indicated by chalk lines. The outside face of the knapsack is a closely woven canvas waterproofed with green paint (the facing colour) while internally a lighter weave of material is used, more like a coarse linen. Dimensions when fastened are 19in by 17in. The 6in diameter central black disc bears the figures '1F' for 1st Fencibles and 'No.5' to identify the company. The inch-wide shoulder straps are sewn top and bottom and adjusted for tightness by means of the breast strap. All buckles are 'D' shaped, without a loop to hold down the free end of the strap. Note the thinness of the side straps. Blankets, if carried, would be folded inside the knapsack rather than rolled on top. (Reconstruction by author)

In undress, and in practice for much of the time on campaign, 'hummel' – plain, featherless – Kilmarnock bonnets were worn by most regiments. (Modern references to 'feathered hummel bonnets' are a contradiction in terms.) At least three regiments, however – the 72nd, 91st and 92nd – wore flat bonnets in undress.

Strictly speaking the purse, or **sporran**, was an accoutrement; but while originally a functional item it had, over the years (like the cavalryman's sabre-tache, which in large measure it resembled), become a largely decorative part of the soldier's uniform and was rarely, if ever, worn on campaign, except perhaps by officers or pipers. In its early form the purse was a simple drawstring leather bag, but by the 1790s it had acquired a large hairy cover at the front. This was generally of goatskin for ORs and spotted sealskin or badger for officers. The drawstrings too became decorative, and about six became the norm, though some units had more, generally arranged in two rows

Regimental colour, 78th Highlanders: light buff with a united plant – not a wreath – in natural colours. The cypher and regimental number are green while the motto appears in gold letters on a red scroll. (Author)

and terminating in large tassels known from their shape as 'bells'.

Pattern and size varied from regiment to regiment, as did the choice of skin. Most regiments opted for a brownish white goatskin, which contrasted well with their dark kilts; but some regiments, for the sake of distinction, opted for black, in which case badger skin rather than sealskin was preferred for the officers' purses. Most regiments, however, expressed their individuality in the number, arrangement and style of the bells and – particularly in the case of officers' purses – the decoration and style of the clasp at the top. Heavy gilt clasps were in favour in the early 1790s, as were the appropriate animal masks on the top flap, but by the 1800s a rectangular leather flap bearing regimental devices was popular. ORs' purses tended, not surprisingly, to be plainer and generally lacked decoration on the top.

The loss of the kilt

Between 1798 and 1806 the five highland regiments which had been serving in India at the outbreak of war straggled home. In every case, having turned over their effectives to regiments remaining in India, they had to be rebuilt around a cadre of officers and NCOs; but found themselves competing for recruits not only with those regiments raised in the 1790s but also with the second battalions authorised in 1803. Not surprisingly, finding sufficient recruits in the highlands was difficult; and the kilt was, rightly or wrongly, identified by Horse Guards (the contemporary term for the army's administrative headquarters) as an impediment to widening catchment areas. The first solution proposed, therefore, was to replace the kilt with tartan pantaloons, or trews as they were now being called. There were useful precedents for this: the 97th Foot had worn pantaloons during its brief existence in the 1790s, as had a number of Fencible regiments, including at least two lowland units. Allan Cameron of Erracht, Colonel of the 79th, was accordingly consulted upon the matter by the Adjutant General's secretary on 13 October 1804:

'I am directed to request that you will state for the information of the Adjutant General, your private opinion as to the expediency of abolishing the kilt in Highland regiments and substituting in lieu thereof the tartan trews, which have been reported to the Commander in Chief, from respectable authority, as an article now become acceptable to your countrymen, easier to be provided, and both calculated to preserve the health, and promote the comfort of the men on service.'

Erracht replied, famously and at length, on 27 October:

'. . . the colonels themselves (those advocating the adoption of trews) being generally unacquainted with the language and habits of Highlanders, while prejudiced in favour of, and accustomed to wear, breeches, consequently adverse to that free congenial circulation of that pure wholesome air (as an exhilarating native bracer) which has hitherto so peculiarly benefited the Highlander for activity and all the other necessary qualities of a soldier, whether for hardship upon scanty fare, readiness in accoutring, or making forced marches, – besides the exclusive advantage, when halted, of drenching his kilt in the next brook as well as washing his limbs and drying both, as it were, by constant fanning, without injury to either, but, on the contrary, feeling clean and comfortable; whilst the buffoon tartan pantaloon, with its fringed frippery (as some mongrel Highlanders would have it), sticking wet and dirty to the skin, is not easily pulled off, and less so to get on again in case of alarm or any other hurry, and all this time absorbing both wet and dirt, followed by rheumatism and fevers, which alternatively make great havoc in hot and cold climates; while it consists with knowledge, that the Highlander in his native garb always appeared more cleanly and maintained better health in both climates, than those who wore even the thick cloth pantaloons. . . . I feel well founded in saying that if anything was wanted to aid the rack-renting Highland landlord in destroying that source which has hitherto proved so fruitful in keeping up Highland corps, it will be that of abolishing their native garb, which His Royal Highness the Commander in Chief and the Adjutant may rest assured will prove a complete death warrant to the recruiting service in that respect; but I sincerely hope His Royal Highness will never acquiesce in so painful and degrading an idea (come from whatever quarter it may) as to strip us of our native garb (admitted hitherto our regimental uniform), and stuff us in a harlequin tartan pantaloon, which, composed of the usual quality that continues as at present worn, useful and becoming for twelve months, will not endure six weeks' fair wear as a pantaloon, and when patched makes a horrible appearance; besides that, the necessary quantity to serve decently throughout the year would become extremely expensive, but above all, take away completely the appearance and conceit of a Highland soldier, in which case I would rather see him stuffed in breeches and abolish the distinction altogether.'

Unfortunately, Horse Guards appear to have taken him at his word. The 79th, ironically enough,

may never have been in any serious danger of losing their kilts, otherwise Erracht might not have been consulted in a 'private' capacity; but in 1809 all five of the 'Indian' regiments, and the 91st besides, not only lost their kilts but were indeed 'stuffed in breeches' and their highland 'distinction' abolished, by the Adjutant General's Memorandum of 7 April:

'As the population of the Highlands of Scotland is found insufficient to supply recruits for the whole of the Highland Corps on the establishment of His

Colonel Hugh Montgomerie (later 12th Earl of Eglinton) by John Singleton Copley. The uniform can be identified as that of the West Lowland Fencibles by the green facings, horizontal pocket flap and a red heart device on the turnback. Although Montgomerie raised the regiment in Ayrshire in 1793 the rank and file wore Government tartan pantaloons and feathered bonnets, while the officers, as shown in this portrait and a caricature by John Kay, wore full highland dress. In the following year Montgomerie also raised the Royal Glasgow Regiment (drafted into the 44th Foot in August 1795), though at its inspection breeches and hats were mentioned rather than quasi-highland dress. (Scottish National Portrait Gallery)

Majesty's Army, and as some of these Corps, laying aside their distinguishing dress, which is objectionable to the natives of South Britain, would in a great measure tend to facilitating the completing of the establishment, as it would be an inducement to the men of the English Militia to extend their service in greater number to these regiments: it is in consequence most humbly submitted for the approbation of His Majesty, that His Majesty's 72nd, 73rd, 74th, 75th, 91st and 94th Regiments should discontinue to

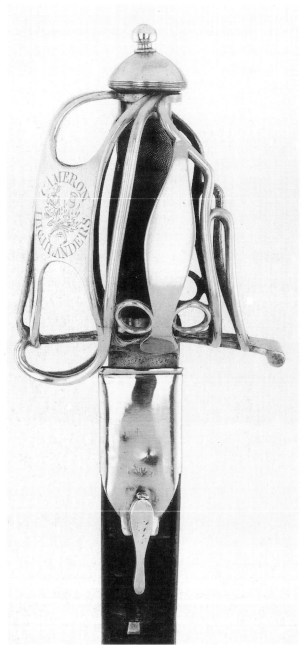

wear in future the dress by which his Majesty's Regiments of Highlanders are distinguished; and that the above Corps should no longer be considered as on that establishment.'

The order was variously received. The 91st, with a very high proportion of highlanders in its ranks, tried as far as possible to ignore it; the 1st Battalion went to Walcheren in trews, and the 2nd Battalion was reported later in the year still to have men wearing kilts. The 75th accepted it with the equanimity to be expected from a regiment which, according to Stewart of Garth, mustered fewer than a hundred men from north of the Tay. It cannot, however, have been received with other than sheer astonishment by the 94th: originally raised in 1793 as the Scotch Brigade and largely recruited in the Edinburgh area, they had never worn highland dress. It is possible that the confusion arose from the fact that they too had served in India; but it seems likelier that their inclusion in the memorandum is a mistake for the youngest of the highland regiments – the 93rd (Sutherland) Highlanders, raised as recently as 1799 and apparently having difficulty in recruiting in its home area. The mistake appears not to have been pointed out. The 93rd certainly had no incentive to do so, quite the reverse; and the 94th probably dismissed it as just another example of incompetence at Horse Guards.

To the six regiments named in the memorandum, of course, should be added the 71st, the oldest of the 'Indian' regiments. Following their return from the Argentinian fiasco in 1807, wearing a mixture of civilian clothing and worn-out uniforms, Col. Pack applied to the Adjutant General's office for permission to adopt tartan pantaloons temporarily, as they could be supplied more quickly. Agreement was forthcoming, and the 1st Battalion fought in them at Vimeiro.

There is, however, more to this than meets the eye. In the first place, in contrast to its modern, knife-pleated counterpart, a box-pleated kilt can be made up remarkably quickly and easily, and Pack's argument on these grounds therefore seems a little odd.

1798 pattern broadsword hilt. This design, in brass, was the first regulation hilt for highland regiments. Previously officers had purchased hilts which suited their own fancy or conformed to a regimental pattern. It was superseded by a steel hilt in 1828. (SUSM)

Moreover, far from the trews being only a temporary expedient, an 1809 inspection report found the 2nd Battalion wearing them too. It is hard to escape the conclusion that Col. Pack was trying to anticipate the threatened 'de-kilting' by establishing the trews, which were apparently acceptable to Horse Guards, as the regiment's uniform, rather than abandoning highland dress entirely as thunderingly advocated by Erracht. In the event he failed, since the 71st, on being converted to light infantry early in 1809, was ordered to adopt breeches. Nevertheless the NCOs and ORs retained their Kilmarnock bonnets, stiffened to resemble shakos; and the pipers, though not without a struggle, were permitted to retain full highland dress.

No sooner were the Napoleonic Wars over than the process began of restoring at least some vestiges of highland dress to these regiments. The 72nd were the first, acquiring feather bonnets and rather striking Royal Stuart tartan trews in 1825; while the 71st regained their trews in 1834, the 73rd appear to have adopted Murray of Atholl trews at least in undress, and the 74th received trews in 1846. The 91st had to wait until 1864, but none of them got their kilts back until 1881.

Weapons

Members of the early highland regiments and independent companies were, famously, armed with a fearsome variety of weapons: pistols, dirk and broadsword as well as the more conventional firelock and bayonet. Such a collection of ironmongery was soon found to be impractical, and as early as May 1775 Col. Stirling of the 42nd reported that his men's broadswords were in store because they had been found too much of an encumbrance. This sensible attitude prevailed and by 1793 swords, as in the rest of the infantry, were carried only by officers, sergeants and musicians. Dirks and pistols were still displayed by officers, but only in full dress.

A surprising variety of swords were carried. Prior to 1798 the only regulation pattern was a rather crude

weapon with an iron Glasgow-style hilt issued to NCOs and to the rank and file. Officers of course purchased their own, generally according to their fancy, although the Breadalbane Fencibles and 116th Highlanders, at least, had a regimental pattern. In 1798 a brass Glasgow-style hilt became the regulation pattern, until replaced by the present steel hilt in 1828. Examples of the 1828 pattern are found with the inner part of the basket cut away to render it more comfortable when riding, but in this period the normal practice was to replace the broadsword with a straight-bladed spadroon, or a sabre of some description. Grenadier officers carried the usual broadsword

Officer of unidentified volunteer or Fencible unit with yellow facings and undifferenced Government sett kilt. There is a black line in the silver lace, while the pink feet and turned- over tops of the stockings show them to be knitted. The dark blue squares in the bonnet dicing are unusual, as are the partly opened lapels. (SUSM)

17

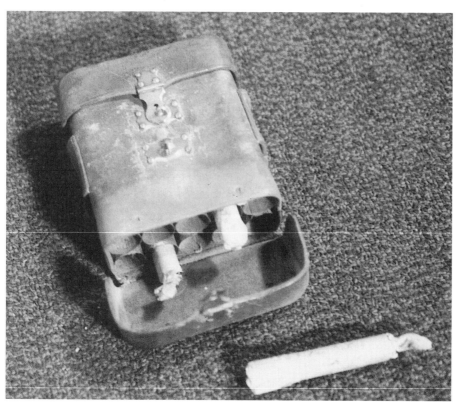

Tin magazine covered in leather and holding twenty rounds of ammunition. This particular example is said to be associated with the Indian Mutiny period, but may in fact be a surviving example of the magazines clipped on to bayonet belts on active service before the introduction of the large cartridge box in the early 1800s. Some highland units wore them as belly boxes, using the sporran belt. (Queen's Own Highlanders)

but Light Company officers normally had sabres, either of the regulation pattern or with a regimental half-basket hilt. There are, however, references to highland Light Company officers carrying dirks instead of swords during the American Revolution, and this may very well have again been the case during the Napoleonic Wars.

The regulation firelock in 1793 was the *Short Land Pattern* .75 calibre, with a 42-inch barrel. The 42nd were certainly equipped with them, and sufficient stocks were probably on hand to equip the newly raised 1/78th and 79th; but the five highland regiments serving in India were already equipped with what would soon become known as the *India Pattern*; also of .75 calibre, its barrel was three inches shorter and the whole weapon was nearly a pound lighter. Originally designed by Stringer Lawrence for the East India Company's troops, it was purchased in large numbers by the Government from 1793 onwards, and in 1797 production of the Short Land Pattern was discontinued in favour of it. The India Pattern has been compared unfavourably by firearms historians with its predecessor; but the shorter barrel went a long way to giving the British soldier an edge

in combat, since the French musket was about the same weight but had a 44-inch barrel. By Waterloo all the highland regiments, with the exception of the 71st who had the *New Land Pattern*, were equipped with India Pattern firelocks.

REGULAR REGIMENTS

1/42nd (Royal Highland)
Raised (R.) 1739 from Independent Companies. In Scotland at outbreak of war. Served (s.) Low Countries 1793-1795; brought up to strength by drafts from 97th, 116th, 132nd and 133rd Highlanders and sent to West Indies 1796-1797; absorbed 79th's effectives before returning, and posted immediately to Gibraltar; Minorca 1798, Egypt 1801, Peninsula 1808-1809, Walcheren, Peninsula 1812-1814, Waterloo 1815.

Recruiting banner, 2/78th, 1804: cream worsted material, 5ft 5in by 5ft, with a fairly crudely painted crown and green scrolls with yellow edge and lettering. Both crown and scrolls are heavily shaded. The essential attractions of enlistment are boldly spelled out to would-be recruits. (Author)

2/42nd

R.1803, s.Peninsula 1810-1812, effectives drafted into 1st Bn. Disbanded (disb.) 1814.

Dark blue facings, 'Jews Harp' bastion loops with red line on outside for ORs, white silk lace for sergeants and gold lace for officers. Kilt: Government sett with red overstripe, at least until c. 1812; brownish white purse. Plain blue Kilmarnock bonnet in undress. Red hackle in mounted bonnet after 1795; Hamilton Smith shows usual all white for grenadiers, but other sources give red over white, and red over green for Light Coy.

1/71st

R.1777 as 73rd but re-numbered 1786. In Madras at outbreak of war. s.Madras and Ceylon (Flank Coys) until 1797. Ireland 1800-1805, Cape of Good Hope and Argentina 1806, Peninsula 1808-1809, Walcheren, Peninsula 1810-1814, Waterloo 1815.

2/71st

R.1803, no foreign service, disb.1814.

Buff facings, square end loops with red line on outside, officers unlaced with silver appointments. Kilt: Government sett with buff and red overstripes until c. 1798, then white and red. Tartan pantaloons instead of kilts 1808; brownish purse. De-kilted 1809, but sergeants and ORs retained stiffened Kilmarnock bonnets.

1/72nd

R.1778 as 78th but re-numbered 1786. In Madras at outbreak of war. s.Madras until 1797; Ireland 1800-1805, Cape of Good Hope 1806-1821.

2/72nd

R.1803, s.Ireland, disb.1816.

Yellow facings, 'Jews Harp' bastion loops with green line on outside, silver lace for officers. Kilt: undifferenced Government sett; white purse. Flat blue bonnets in undress. De-kilted 1809.

1/73rd

R.1780 as 2/42 but re-numbered 1786. In Madras at outbreak of war; s.Madras until 1805, New South Wales 1809-1816.

2/73rd

R.1809, s.Germany and Holland 1813-1814, Waterloo 1815. disb.1816.

Dark green facings, 'Jews Harp' bastion loops with red line on inside, gold lace for officers, Kilt: Government sett with red overstripe; brown purse. De-kilted 1809.

74th (Argyle)

R.1787 for service in India. s.Madras until 1805, most notably at Assaye; Walcheren 1809, Peninsula 1811-1814.

White facings, square end loops with red line on outside, gold lace for officers. Kilt: Government sett may have been undifferenced though ordered by Adjutant General to be distinguished by white overstripe when permitted to resume highland uniform in 1846; white purse. De-kilted 1809.

75th

R.1787 for service in India. s.Madras until 1805, slow to recruit and did not go abroad again until 1811 when sent to Sicily, and to Corfu in 1814.

God save great George our King

An Old PERFORMER playing on a New INSTRUMENT.
or one of the 42d Touching the Invincible.

Yellow facings, square end loops in pairs, two yellow and one red line in lace, officers lace silver. Kilt: undifferenced Government sett. De-kilted 1809.

1/78th (Ross-shire Buffs)

R.1793, s.Low Countries 1794-1795, Cape of Good Hope 1796, India 1797-1811, Java 1811-1816.

2/78th (a)

R.1794, s.Cape of Good Hope 1795 but drafted into 1st Bn June 1796.

2/78th (b)

R.1804, s.Shorncliffe, training as light infantry; Gibraltar 1805, Sicily 1806 (fought at Maida), Egypt 1807; draft for India diverted to Walcheren, not going out to 1st Bn. until 1810. Holland and Belgium 1813-1816.

Pale buff facings, 'flower pot' bastion loops with green line on outside, officers unlaced with gold appointments. Kilt: Government sett with white and red overstripes; white purse. Blue Kilmarnock bonnet with pale buff band worn in undress.

1/79th (Cameron)

R.1793, s.Low Countries 1794-1795, Martinique 1795-1797, Guernsey 1798, Helder 1799, Egypt 1801, Copenhagen 1807, Peninsula 1808-1809, Walcheren, Peninsula 1810–1814, Waterloo 1815.

2/79th

R.1804, no foreign service, disb.1814.

Green facings, square end loops in pairs with two red lines and a central yellow line in lace, gold lace for officers. Kilt: Cameron of Erracht sett; white over black purse. Finart shows Kilmarnock bonnet with diced band in undress.

1/91st (Argyllshire)

R.1794 as 98th but re-numbered 1798. s.Cape of Good Hope 1795–1803, Hannover 1805, Peninsula 1808–1809, Walcheren, Peninsula 1812–1814, Belgium and France 1815.

2/91st

R.1803, s.Germany and Holland 1814 (Bergenopzoom), disb.1814.

Yellow facings, square end loops in pairs with black dart on outside and black line on inside of lace, officers unlaced with silver appointments. Kilt: probably undifferenced Government sett. Flat blue bonnet in undress. De-kilted 1809/10.

1/92nd (Gordon)

R.1794 as 100th, re-numbered 1798. s.Gibraltar and Corsica 1795-1797, Ireland 1798, Helder 1799, Egypt 1801, Copenhagen 1807, Peninsula 1808-1809, Walcheren, Peninsula 1810-1814, Waterloo 1815.

Regimental colour, 71st Highlanders, 1805-1806; ochre buff with red cartouche and an unusually thick wreath, depicted naturally. This colour was surrendered at Buenos Aires in 1806. (Author)

2/92nd
R.1803, s.Ireland, disb.1814.

Light yellow facings, square end loops in pairs with blue line on outside, silver lace with black line for officers. Kilt: Government sett with yellow overstripe; white purse. Flat blue bonnet in undress.

1/93rd (Sutherland)
R.1799, s.Cape of Good Hope 1806-1814, America (New Orleans) 1815.

2/93rd
R.1813, s.Newfoundland 1814-1815, disb.1815.

Yellow facings, 'Coldstream' loops in pairs with yellow line on outside, silver lace for officers. Kilt: undifferenced Government sett; black purse; tartan pantaloons and hummel bonnets worn at New Orleans. 'Sutherland' dicing on Kilmarnock bonnets.

97th (Inverness-shire)
R.1794, s.Guernsey but later put aboard the Channel Fleet to serve as marines. Drafted August 1795; both flank companies to 42nd, battalion companies to Marines.

Green facings, square end loops in pairs, probably green and red lines in lace, gold lace for officers. Kilt: Government sett with red overstripe; tartan pantaloons normally worn instead of kilts; white purse. Plain blue Kilmarnock bonnets in most orders of dress. Round hats, red jackets, and white trousers worn while serving with the fleet.

109th (Aberdeenshire)
R.1794, s.Channel Islands, and allocated to 'descent' on West Indies 1795, but instead drafted into 53rd in August.

Yellow facings, square end loops, silver lace for officers. Kilt: probably undifferenced Government sett; white purse.

116th (Perthshire)
R.1794, s.Ireland but drafted into the 42nd in August 1795.

White facings, square end loops in pairs, silver lace with black border for officers. Kilt: Government sett with double yellow overstripe; black purse.

132nd
R.1794 but probably never completed before being drafted into 42nd, August 1795.

Black facings?

133rd (Inverness Volunteers)
R.1794 but probably never completed before being drafted into 42nd, August 1795.

Black facings?

FENCIBLES

In addition to the line regiments raised for unlimited service, a number of Fencible regiments were recruited in the 1790s. Scotland had at that time no militia and Fencible units, required to serve in a home defence capacity only, were seen as filling the gap. Accordingly seven were raised in 1793, five of them highland units. Although the same bounties and rates of pay were extended to them the men were ineligible for pensions at the conclusion of their service, and the

An apparently accurate depiction of Piper Clarke and the 71st at Vimeiro, by Atkinson. Note the tartan pantaloons and feather bonnets worn by the rank and file, and Clarke's grenadier company wings and hackle. (SUSM)

officers would not draw half pay, nor could they exchange with officers in line regiments. Officers who did wish to enter regular service normally did so by first purchasing an ensign's commission in an Independent Company and proceeding from there.

The limitations of their terms of service soon became apparent when they refused to volunteer for service south of the border, and the Strathspey Fencibles actually mutinied. The Gordon Fencibles were eventually prevailed upon to volunteer; but when further Fencible regiments were raised in 1794 and 1795 their terms of service were enlarged, first to allow for their being sent anywhere in Britain or Ireland, and then in 1798/9 to anywhere in Europe. In return for these wider terms of service the officers of these new regiments became eligible for half pay and their men for pensions if wounded.

All Fencible regiments were disbanded at the Peace of Amiens in 1802; and when war broke out again shortly afterwards the War Office decided, sensibly, to concentrate on raising second battalions for line regiments.

The Regiments

NB: Uniform details below are based on Hamilton Smith's chart. Unless otherwise noted all regiments had kilts, diced hose and feather bonnets.

Aberdeen (Princess of Wales)
R.1794, s.Ireland. Lemon yellow facings, square end loops, silver lace for officers. Kilt: undifferenced Government sett; white purse.

Angus
R.1794. Yellow facings. Kilt: undifferenced Government sett (trews also recorded); white purse.

Western (Argyle)
R. 1793 by Marquis of Lorne for service in Scotland only. Disb.1799. Pale yellow facings, square end loops, silver lace for officers. Kilt: undifferenced Government sett; brownish purse.

1st Argyle
R.1794 by Col. Henry Clavering, s.Ireland. Blue facings and fringe on shoulder straps. Kilt: undifferenced Government sett; white purse.

2nd Argyle
R.1799 by Col. Archibald McNeil of Colonsay,

s.Gibraltar. Yellow facings, and fringe on shoulder straps. Kilt: undifferenced Government sett; white purse.

(NB: The above appear to have been three quite separate units, not three battalions of one regiment.)

Banffshire (Duke of York's)
R.1799. Blue facings. Kilt: undifferenced Government sett; white purse.

1st & 2nd Bns. Breadalbane
R.1793 for service in Scotland only. Disb.1799. White facings? (Hamilton Smith gives yellow but surviving coat appears to have white), silver lace with black border for officers. Kilt: Government sett with double yellow overstripe; black purse.

3rd (Glenorchy) Bn. Breadalbane
R.1794, s.Ireland. Uniform details probably as above though Hamilton Smith gives a red overstripe on tartan.

Caithness Legion
R.1794, s.Ireland. Yellow facings, silver lace for officers, tartan pantaloons – probably undifferenced Government sett; white purse.

Dumbarton
R.1794, s.Guernsey and in Ireland 1798. Black facings. Kilt: undifferenced Government sett (trews also recorded). Bonnet dicing red and black, on white. Hose: grey/black check on white.

Lord Elgin
R.1794. Green facings. Kilt: undifferenced Government sett (trews also recorded); white purse.

Fraser
R.1794, s.Ireland 1798. Black facings. Kilt: Fraser tartan - basically red sett.

Glengarry
R.1794, s.Jersey and Guernsey. Yellow facings. Kilt: probably undifferenced Government sett, though perhaps Glengarry sett; white purse. Sutherland dicing on bonnet.

Gordon (Northern)
R.1793 for service in Scotland, later extended to England. Disb.1798. Lemon yellow facings, square end loops in pairs, silver lace for officers. Kilt: Government sett with yellow overstripe; white purse. Black leather equipment.

1: Ensign Hon. Arthur Wesley, 73rd Highlanders, 1787
2 & 3: Officer and Private, 42nd Royal Highland Regiment 1792

A

1: Recruiting Sergeant, 71st Highland Regiment, 1793
2: Private, Light Company, 78th Highlanders, 1793
3: Grenadier, 79th Cameron Volunteers, 1793

B

1: Private, 1st or Strathspey Fencibles, 1790s
2: Officer, 1st or Strathspey Fencibles, 1790s
3: Private, Rothesay and Caithness Fencibles, 1794

1: Private, 97th Inverness-shire Regiment, 1794
2: Field officer, 109th Aberdeenshire Regiment, 1794
3: Battalion Company, 100th (Gordon) Highlanders, 1794

D

1: Officer, 116th Highlanders, 1795
2: Pioneer, 93rd (Sutherland) Highlanders, 1800
3: Light Company, Dumbarton Fencibles, 1798

E

Pipers
1: Pipers, Reay Fencibles, 1794
2: Piper, 1/71st (Glasgow) Highlanders, 1806
3: Piper, Light Company, 42nd Royal Highland Regiment, 1812

1: Battalion Company, 91st Highlanders, 1810
2: Lieutenant, 2/78 Highlanders, 1806
3: Battalion Company, 79th (Cameron) Highlanders, 1815

G

92nd (Gordon) Highlanders, 1815
1: Grenadier, 92nd service dress
2: Officer, undress
3: Sergeant-Major, full dress

H

Loyal Inverness

R.1795, s.Ireland with sufficient distinction to be re-titled 'Duke of York's Royal Inverness-shire Highlanders'. Yellow/buff facings, changed to blue 1798, silver lace for officers. Kilt: Baillie tartan - basically light blue sett.

Regiment of the Isles

R.1799. Yellow facings, gold lace for officers. Kilt: probably MacDonald tartan.

Lochaber

R.1799, s.Ireland. Black facings. Kilt: undifferenced Government sett; white purse. White hackle in bonnet.

McLeod (Princess Charlotte of Wales')

R.1799, s.Ireland. Yellow facings. Kilt: undifferenced Government sett White hackle in bonnet.

Perthshire

R.1794 (never completed?). Yellow facings. Kilt: undifferenced Government sett; white purse.

Reay

R.1794, s.Ireland 1798. Pale blue facings, silver lace in pairs for officers. Kilt: Mackay tartan - basically green sett; badger purses for officers and sergeants, white goatskin for rank and file.

Ross-shire

R.1796 (two companies only). Yellow facings. Kilt: Government sett, probably with red overstripe.

Ross and Cromarty Rangers

R.1799. Yellow facings. Kilt: Government sett with red overstripe.

1st Bn. Rothesay and Caithness

R.1793 for service in Scotland only. Disb.1799. Yellow facings, square end loops, blue line in lace, silver lace for officers. Gordon tartan pantaloons; white purse. Bonnet dicing red and yellow on white, red, yellow and white hackle.

2nd Bn. Rothesay and Caithness

R.1795, s.Ireland. Uniform as above, red waistcoat may have been worn as 2nd Bn. distinction.

Highlanders, by Atkinson, 1807. They are identifiable as members of the 73rd by the bastion lace and green facings. The kilts have the red overstripe inherited from the 42nd. Note what appears to be a tin magazine lying beside the brown envelope knapsack in the foreground. The brown knapsack has a light coloured circle in the centre with what may be an elephant underneath, and 'No.3' – presumably a company number – above. (Queen's Own Highlanders)

Royal Clan Alpine

R.1799, s.Ireland. Blue facings and fringe on shoulder straps. Kilt: undifferenced Government sett.

Strathspey (Grant)

R.1793 for service in Scotland only. Disb.1799. Green facings, square end loops in pairs, gold lace for officers. Kilt: Government sett with red overstripe; badger purses for officers, white goatskin for NCOs and men.

Sutherland

R.1793 for service in Scotland only, but volunteered in 1797 for service in Ireland. Disb.1799. Yellow facings, 'flower pot' bastion loops, silver lace for officers. Kilt: undifferenced Government sett; black purse. Sutherland dicing on bonnet.

THE PLATES

A1: Ensign Hon. Arthur Wesley, 73rd Highlanders, 1787

The future Duke of Wellington, then the Honourable Arthur Wesley, began his military career on 7 March 1787 when he was gazetted ensign in the 73rd Highland Regiment. Since they were in India at the time all his service with the regiment will have been spent at the depot. He wears a coat conforming to the 1768 Regulations, with green facings and gold lace. His belted plaid is the usual Government or Black Watch sett. The 73rd had originally been the 2/42nd and it appears to have retained a number of features of its parent regiment's uniform, including the addition of a red overstripe to the sett used for the little kilt.

A2 & A3: Officer and Private, 42nd Royal Highland Regiment, 1792

Based on watercolours by Dayes, the officer wears full dress uniform and a broadsword with an elaborate Stirling-made hilt, conforming to no official pattern. Since the 42nd traditionally drew its officers from Perthshire and other areas east of the Great Glen they tended to buy their broadswords in Stirling rather than Glasgow. His belted plaid, probably with pre-sewn pleats, is of the plain Government sett. The battalion company private, on the other hand, wears a little kilt of the Government sett with red overstripe. This overstripe is usually said to have been worn only as a distinction by the Grenadier company; but Col. David Stewart of Garth, who joined the 42nd in 1787 and served with the regiment until 1803, states categorically that it was used throughout the regiment for the kilt as distinct from the plaid. Up until the American War it appeared on both belted plaids and kilts, since the latter were originally made up from worn-out plaids. By the 1780s, however, the kilt had become the normal order of dress, and as such needed to be made from new rather than old and shabby material, and the distinction then seems to have been made between kilt and plaid. A notable feature of all tartan setts worn in the 1790s and early 1800s is that the actual size of the checks was rather smaller than at the present time. The hose as depicted by Dayes are knitted ones, easily identifiable by the turned-over top, and consequently lack the black edging to the check associated with the regiment. An inspection report of 8 June 1791 noted that the appearance of the regiment was improved 'by having now white accoutrements instead of black ones'.

B1: Recruiting Sergeant, 71st Highland Regiment, 1793

The 71st was at this time in India and this sergeant therefore appears as he would have done when recruiting at home, dressed in full highland uniform

Major James McDonnell of Glengarry, 2/78th, c.1805. McDonnell joined the newly raised battalion from the 5th Foot in 1804, fought at Maida and was promoted to lieutenant-colonel in 1809, but transferred to the Coldstream Guards in the following year. Afterwards he distinguished himself in the defence of Hougoumont. Note the McKenzie tartan plaid and the fully buttoned-over coat. (Private collection)

71st Highlanders at Waterloo, after Captain Jones; they wear the Light Infantry style uniform adopted in 1809. Note that the officer has a true light infantry shako rather than a stiffened bonnet. The inclusion of a colour in this scene is erroneous since the 71st had none at Waterloo: the set which replaced those lost at Buenos Aires had themselves been mislaid after a dinner at Carlton House to celebrate Wellington's return from the Peninsular War.... Note that the piper, presumably an unauthorised one belonging to a battalion company, is wearing Light Infantry uniform rather than highland dress.

including a belted plaid with distinctive buff and red overstripes. The rosette in his bonnet is a bunch of regimental lace, and like all recruiters he cultivates a rather flashy appearance, including an officer's sash. Recruiting parties tended to be forgotten when new issues of clothing were made to regiments overseas, however, and the black sword belt shown would by 1793 have been obsolescent (although Hamilton Smith shows the Gordon Fencibles with black equipment). In India the regiment wore what was known as 'East India Uniform'; a short single-breasted red coat, round hat and white gaiter trousers, although highland dress seems to have continued to have been worn on occasions, especially by pipers: indeed, one print appears to depict a piper of the 73rd rather incongruously wearing gaiter trousers under his belted plaid and diced hose.

B2: Private, Light Company, 78th Highlanders, 1793

The 78th was the first regiment to be raised at the outset of the Revolutionary War. Like the 71st it had buff facings but in this case they were a very pale shade, to all intents and purposes an off-white. The tartan chosen also resembled that worn by the 71st but was distinguished by white rather than buff overstripes. Prior to the substitution of 'soft' for 'hard' tartans in the second half of the 19th century these overstripes were woven in using silk thread in order to produce a suitable contrast. Correspondence belonging to the Bannockburn firm, William Wilson & Co., reveals that the 78th tended to be rather 'vexatious' and difficult to please in this regard, possibly because the white overstripe may originally have been intended to be a very pale buff corresponding to the facings. Unlike the Black Watch, Gordon and A&SH setts, the size of the 78th checks has not increased since the 1790s. The knapsack is the usual canvas style, painted by way of waterproofing in the regimental facing colour. The 78th appear to have retained their buff knapsacks as late as February 1821. Leather caps with peaks for the light company

Regimental colour, 2/92nd: lemon yellow with silver star in the centre and blue scrolls. It is distinguished from the 1st Battalion colour by the broad red band across the centre. The 2nd Battalion was disbanded in 1814 but it is just possible that this colour was carried by the 1st Battalion at Waterloo since their own colours were in an extremely tattered state after service in the Peninsula. (Author)

are mentioned, but there is no indication as to the pattern used. Highland flank companies wore bonnets in most orders of dress, however, reserving the caps for formal occasions.

B3: Grenadier, 79th Cameron Volunteers, 1793

The most striking feature of the 79th uniform was its unique tartan. Unlike those worn by the other regular highland units it was not based upon the Government sett but was instead specially designed for the regiment by the commanding officer's mother, introducing a yellow overstripe to an existing local sett. While the latter is now claimed as a MacDonald sett it may not have been recognised as such at the time. As has been the case with most of the regimental tartans it was originally rather lighter than modern versions. The reconstruction here is based upon a piece of tartan woven to the original specification.

Grenadier caps were still the prescribed wear for formal occasions but bonnets were frequently worn

for preference, distinguished not merely by the colour of the cut feather hackle but also by the coloured tourie or woollen ball on the crown: white for grenadiers, red for battalion companies and green for light company men.

C1: Private, 1st or Strathspey Fencibles, 1790s

Raised by Sir James Grant of Grant in 1793, the Strathspey Fencibles spent their entire existence in Scotland and were notable for mutinying twice in two years. The regiment, not having volunteered for service outside Scotland, was disbanded in 1799.

The 1st Fencibles appear to have been distinguished by a red overstripe in their tartan. This was certainly specified for the drummers, and a hand-coloured version of Kay's print depicting Sir James and his regiment also shows a red overstripe. Members of the Light Company had the usual lace wings and wore plain blue Kilmarnock bonnets.

C2: Officer, 1st or Strathspey Fencibles, 1790s

Based upon a sketch by John Kay of Capt. John Rose of Holme, the uniform illustrated served not only as an undress, but was also worn by officers of highland regiments serving in the Indies. A miniature of an officer of the 79th, in 1795, shows a very similar round hat with bearskin crest, and three gold bands – either thin lace or chains – and there is a comment that officers of the 42nd, also serving in the Caribbean at the time, were wearing the same pattern hat that year. With highland dress a basket-hilted broadsword would be worn but in undress the lighter and less cumbersome spadroon was obviously preferred.

C3: Private, Rothesay and Caithness Fencibles, 1794

This unit was raised by Sir John Sinclair of Ulbster, like Sir James Grant a keen improver of agriculture and anything else which took his fancy. The odd title is accounted for by the fact that the two areas were paired off for the purposes of Parliamentary representation. One of Sir John's enthusiasms was highland dress, and he clothed his regiment in tartan trousers on the grounds that they were of greater antiquity than kilt and plaid. The bonnet is unusual in that the green squares at the intersection of the red checks are replaced with yellow ones. The yellow overstripe in the tartan is presumably also derived from the facings. The breastplate bears the inscription 'FLODDEN' over the usual thistle, a none too happy reminder of the last time Caithness men were said to have been mustered for war. The first battalion saw out its service in Scotland, though finally forming part of the garrison of Berwick; but a second battalion raised in 1795 served in Ireland during the 1798 rebellion and was disbanded in 1802. Hamilton Smith shows red waistcoats for this regiment, which may have been a 2nd Bn. distinction since Sinclair himself is depicted in his portrait wearing the usual white one.

D1: Private, 97th Inverness-shire Regiment, 1794

Sir James Grant also raised a marching regiment, uniformed similarly to his Fencibles. It is possible that the 97th may have worn the usual Government sett since a piece of this tartan, without a red overstripe, was found in one of the surviving 97th knapsacks at Castle Grant (now in the Seafield Collection at Fort George). There is evidence, however, that these knapsacks were later used by local militia. Tartan pantaloons or trousers were worn in most orders of dress, and the plain blue Kilmarnock bonnet appears to have been worn by all the companies.

D2: Field officer, 109th Aberdeenshire Regiment, 1794

On 8 March 1794 Alexander Leith-Hay, a half-pay officer, wrote to the War Office: 'In the beginning of the war I offered to raise a regiment of Highlanders and will still undertake to raise one on the same establishment and terms with those now going on in Scotland.' His offer was accepted this time, and letters of service were granted on 2 April, much to the dismay of the recruiters for the 100th (Gordon)

Highlanders. It was raised in the lowlands of eastern Aberdeenshire, and the muster rolls reveal only 33 'Macs' out of 694 NCOs and men. Nevertheless, like many other regiments raised in the area before and since, it was carried on the highland establishment, as confirmed by a letter from Donaldson, the regimental agent in January 1795, referring to the 109th (Highland) Regiment.

This field officer, distinguished by riding breeches and boots and a spadroon, has the lemon yellow facings common to Aberdeenshire regiments. Leith-Hay's brother, James Leith, raised the Aberdeen (Princess of Wales) Fencibles shortly afterwards, and Hamilton Smith shows them to have had the usual Government sett kilt and a white goatskin sporran as well as lemon yellow facings. The 109th presumably wore a very similar uniform.

Print by Finart apparently depicting a grenadier of the 92nd. As it was normal to hand-colour this and similar prints to depict all three highland regiments in Paris, the presence of a peak to the bonnet cannot be relied upon as evidence that a prohibition in regimental orders was being flouted. (SUSM)

D3: Battalion company, 100th (Gordon) Highlanders, 1794

Although letters of service for raising the 100th (later 92nd) were granted as early as 10 February 1794, the recruiters for the 109th were so successful in eastern Aberdeenshire that the Marquis of Huntly, nominated by his father to command the regiment, had for the most part to look further afield for his men. Only 124 of the original recruits came from Aberdeenshire, with a further 82 from Banff, most of both groups coming from his own lands in Strathbogie; no fewer than 240, by contrast, came from Inverness-shire, despite the efforts of rival recruiters for other highland regiments.

The tartan – the Government sett with yellow overstripe – was devised in 1793 for the Northern Fencibles, raised by Huntly's father, the Duke of Gordon. On 15 April 1793 William Forsyth, a tartan manufacturer and clothier who handled contracts for most of the regiments raised in the area, wrote to the Marquis:

'When I had the honour of communing with His Grace the Duke of Gordon, he was desirous to have patterns of the 42nd Regiment plaid with a small yellow stripe properly placed. Enclosed three patterns of the 42nd plaid all having yellow stripes. From these I hope his Grace will fix on some of the three stripes. When the plaids are worn the yellow stripes will be square and regular. I imagine the yellow stripes will appear very lively.'

While the three original samples have not survived, the Duke evidently settled upon the second of them. Surviving kilts of the period in the regimental museum, SUSM and NAM reveal that although the sett is the same one worn today, with a single yellow overstripe, the checks were woven very much smaller and the colours lighter.

When first raised the Gordons had goatskin knapsacks, costing 1s. 6d. more than the Government allowance; but in 1796 they changed to the more usual canvas ones, painted yellow with a red roundel bearing a crown and thistle and the words 'Gordon Highlanders'.

E1: Officer, 116th Highlanders, 1795

In 1793 the Earl of Breadalbane raised a Fencible regiment in Perthshire which eventually boasted

92nd and Scots Greys charging together at Waterloo, after Captain Jones. Note the height of the Kilmarnock bonnets underneath the feathers.

three battalions, and like some of the other noble personages behind the raising of Fencible corps in that year his thoughts soon turned to a marching regiment. The result was the 116th (Perthshire) Highlanders, disbanded in Dublin in 1795. The first and second Fencible battalions were raised for service in Scotland only and were disbanded in 1798. However, the third or Glenorchy Battalion served in Ireland from 1795 to 1802.

The uniform shown was worn by both the officers of the Breadalbane Fencibles and those of the 116th. Hamilton Smith gives the Fencibles yellow facings and notes a red overstripe in the sett worn by the third battalion. In actual fact both regiments had the distinctive double yellow overstripe on their kilts, and a Fencible officer's coat in the SUSM appears to have white facings and a black line in the silver lace. The sword hilt is equally distinctive. Both regiments had a virtually identical pattern.

E2: Pioneer, 93rd (Sutherland) Highlanders, 1800

Pioneers were identified by their axes and other tools and a tan leather apron. A fur grenadier-style cap with a red enamelled plate was the prescribed headdress though, like the Grenadiers and Light Infantry, on most occasions they wore bonnets. The most distinctive feature of the 93rd's uniform is the dicing on the bonnet. This particular pattern, featuring white squares instead of green at the intersection of the red checks, is known as 'Sutherland dicing' and was traditionally used by regiments raised in that area. It was, however, sometimes difficult to obtain. On 10 January 1810 Major Dale complained to the suppliers, William Wilson & Sons: 'The bonnets last sent are not the Regimental Pattern ... the 93 Regl Bonnets are three rows of plain Red & White dice, without any green whatever.' Interestingly, a Sutherland Local Militia bonnet in the SUSM also has the ordinary red and green dicing rather than the Sutherland dicing one would expect.

E3: Light Company, Dumbarton Fencibles, 1798

Letters of service were granted to Campbell of Stonefield on 11 October 1794 to raise this regiment, for service anywhere in the British Isles. In the summer of 1795 they were sent to Guernsey, and

Officer, 42nd: most of the items of clothing and equipment are correct for the 1815 period, although overalls would have been worn on campaign in place of kilt and sporran, and the coat would be buttoned over. The kilt itself probably dates to the 1820s, but usefully shows how very much smaller and lighter in colour the Government sett was at this time. The stockings are modern replacements and the black line did not normally appear on knitted ones. (SUSM)

39

Regimental colour, 93rd Highlanders: yellow with a thistle and crown device in the centre instead of the cartouche common to line regiments. This colour was carried at New Orleans. (Author)

while there reduced to an establishment of only 500 men – a move which according to Stewart of Garth gave a timely opportunity to weed out all the troublemakers (chiefly Glaswegians). In 1797 they were posted to Ireland, serving in the Dublin garrison, though after the outbreak of the republican rebellion in the following year they became Gen. Lake's bodyguard and later fought under Sir John Moore. A detachment also fought in the crucial battle of Arklow, while after the rebellion another detachment had the unusual job of guarding 400 rebel prisoners shipped to Prussia.

The rather sombre uniform is recorded in Hamilton Smith's Fencible chart. Black facings were not unknown, but the grey/black dicing on the hose appears to be unique to this regiment.

F: Pipers

The status of pipers in highland regiments was by no means clear-cut. While it is usually stated that none were authorised until January 1854 this is not entirely true. Two fifers were carried on the establishment of line regiments; and in the letters of service granted to

a number of highland regiments, including e.g. the 79th and 97th Foot and the Strathspey and the Northern Fencibles, pipers were substituted for the fifers. In other regiments pipers may simply have been enlisted as fifers; and a rather crude print depicting a piper, probably of the 73rd, actually shows a fife case slung by his side. Otherwise pipers might be carried on the books as private soldiers, officers' servants, or even as drummers and bandsmen. In consequence there were no regulations covering their dress.

F1: Piper, Reay Fencibles, 1794

This was one of the Highland Fencible units posted to Ireland in 1795, and subsequently served with some distinction against the rebels in 1798, most notably perhaps at the Hill of Tara on 26 May when three companies led by Capt. Hector MacLean routed a large army of United Irishmen virtually unaided.

The light blue facings were quite distinctive, and like a number of Fencible units a 'local' tartan was worn – in this case Mackay – in preference to the Government sett. Other recorded or alleged variations included the Baillie tartan (a blue sett) for the Inverness-shire Fencibles, and Fraser and McDonnell of Glengarry for the Fraser and Glengarry Fencibles respectively. The pipe banner is based upon a surviving example, and the silver lace indicates that he is the pipe-major.

F2: Piper, 1/71st (Glasgow) Highlanders, 1806

By the 1800s the 71st had changed the buff overstripes in their tartan to white, probably on resuming highland dress on their return from India in 1798. The sett was now therefore identical to that worn by the 78th, but a distinction seems to have been maintained by arranging the kilt differently (see **G2**). The pipe-banner illustrated was captured at Buenos Aires in 1806; a second one, also lost there, was similar but had a crimson background. In 1808 the 71st switched to tartan pantaloons instead of kilts, and had to adopt Light Infantry uniform in the following year, although their pipers were permitted to retain full highland dress. This concession apparently only extended to the two pipers authorised for the Grenadier Company, since although Atkinson shows Piper Clark wearing a kilt at Vimeiro, Capt.

Jones depicts a piper, presumably belonging to a battalion company, wearing ordinary Light Infantry uniform at Waterloo.

F3: Piper, Light Company, 42nd Royal Highland Regiment, 1812

The 42nd wore plain blue Kilmarnock bonnets for undress, and the piper is identified as belonging to the Light Company by the regimental pattern half-basket hilted sabre. The red overstripe in the tartan was certainly used as late as 1812 but may have been discontinued by 1815, possibly as a result of the difficulty in keeping the regiment supplied with tartan during the Peninsular campaigns. The overstripe is, however, still shown in an Atkinson print dated 4 June 1816, purporting to show the 42nd and the Scots Greys at Waterloo. Hamilton Smith shows a square breastplate for the 42nd, but surviving examples are oval.

G1: Battalion Company, 91st Highlanders, 1810

Raised in 1794 as the 98th (Argyllshire) Regiment and renumbered as the 91st in 1798, this was one of the highland regiments ordered (inexplicably in this case) in 1809 to relinquish the kilt. The order to do so, however, arrived after that year's allotment of tartan had been issued, so permission was granted to make the material up into trews; and these were worn at Walcheren, together with the flat bonnets normally reserved for undress. A little uncertainty exists as to the tartan sett actually worn by the regiment. It was described in 1794 as being green and black, which might be something not unlike the modern Mackay sett, but it was most likely the Government sett as depicted here.

G2: Lieutenant, 2/78th Highlanders, 1806

This figure is based on a portrait of Lt. William McKenzie, then the Adjutant of the 2nd Battalion. Particularly noteworthy is the practice, common

Death of Colonel Macara of the 42nd at Quatre Bras, after Jones. Badly wounded, he was being carried from the field when some French lancers murdered him together with his carrying party.

'Roast Chestnuts', a mild example of a series of sometimes viciously pornographic caricatures of highland soldiers in Paris after Waterloo, in this case coloured to represent a member of the 79th. Note the bonnet tightening tapes. (SUSM)

amongst highland officers, of fully buttoning over the lapels of the coat. This avoided any problems either with the shoulder plaid, or indeed with the sash, which in contrast to the practice in line regiments was worn over the shoulder rather than tied around the waist. Shortly after this portrait was painted the 78th adopted an elephant badge on both breastplate and bonnet, commemorating the 1st Battalion's part in the battle of Assaye. He wears the 'McKenzie' sett, distinguished from that worn by the 71st by the centrally placed red vertical overstripe. The 78th appear not to have adopted knitted hose until 1848.

G3: Battalion Company, 79th (Cameron) Highlanders, 1815

Based upon a Finart print, this figure illustrates a typical undress uniform, with tartan trousers substituted for the kilt and a plain or hummel Kilmarnock bonnet. Soldiers walking out were required to carry their bayonets at all times in order to distinguish them from 'vile mechanics'. The original Finart plate is a good example of the problems which a too close reliance upon such sources can cause. Although identifiable as a Cameron by the double square ended lace loops and the Kilmarnock bonnet (the Gordons

also had double loops but wore flat bonnets), the figure is coloured to represent a member of the 42nd with blue facings and a greenish tartan.

An interesting selection of figures from all three highland regiments in Paris; the bandmaster's legwear (second right), if correctly depicted, is interesting. (SUSM)

H: 92nd (Gordon) Highlanders, 1815

The three figures in this plate are intended to illustrate the different uniforms worn by members of the regiment during the 1815 campaign. After suffering heavy casualties at Quatre Bras on 16 June they were involved, early in the battle of Waterloo, in one of its most famous incidents: the charge of the Gordons and the Scots Greys against D'Erlon's corps, described here by Lt. Robert Winchester, one of those actually taking part in the attack:

'About two or three o'clock in the afternoon a Column between 3,000 to 4,000 men advanced to the hedge at the roadside.... Previous to this the 92nd had been lying down under cover of the position when they were immediately ordered to stand to their arms, Major General Sir Denis Pack calling out at the same time, "92nd, everything has given way on your right and left and you must charge this Column,"

upon which he ordered four deep to be formed and closed in to the centre. The Regiment, which was then within about 20 yards of the Column, fired a volley into them.... The Scots Greys came up at this moment, and doubling round our flanks and through our centre where openings were made for them, both regiments charged together, calling out "Scotland for ever", and the Scots Greys actually walked over this column, and in less than three minutes it was totally destroyed.'

H1: Grenadier, 92nd, service dress

This figure can be taken as representative of all four kilted regiments (42nd, 78th, 79th and 92nd) in the Low Countries in 1815. Apart from the usual regimental distinctions, however, the 92nd differed from the others in being forbidden by regimental standing orders to wear the leather peak on their

▲ A good print by Vernet depicting a soldier of the 79th and his family. Other versions exist coloured as soldiers of the 42nd and 92nd. The usual allowance of wives was six to a company, though it is clear from Anton's memoirs that no restrictions were placed on sergeants' wives. (SUSM)

▶ A field officer of a highland regiment, apparently taking exception to his English colleague's taste in coats. (SUSM)

bonnets. A Finart print of 1815 depicting a grenadier of the 92nd does in fact show this feature, but given the propensity of the publishers to colour the prints to taste it cannot be regarded as reliable evidence that the order was in fact flouted.

Hamilton Smith shows a square breastplate for the 92nd but surviving examples from the period are oval. The blankets issued to the brigade for this campaign were, according to Sgt. Anton, unusually large: 'In Ghent all our great-coats were taken from us, and in place of them we received blankets; these,

for quality and size, were such as an army had never received before, and such as were worthy of England to give. They were intended not only to be our covering in camp, but, upon any emergency, to serve for a tent also. We had them all looped and prepared for this purpose.'

H2: Officer, undress
This uniform has been reconstructed from surviving garments belonging to Lt. Claude Alexander (commissioned 19 September 1805) and Ensign John

Bramwell (commissioned 29 July 1813), in the SUSM and NAM respectively. Bramwell was wounded at Quatre Bras, but Alexander, the Adjutant, came through unscathed. The red Light Company waistcoat was worn by Alexander at the Duchess of Richmond's ball. The black line in the silver lace appears to have been worn since the regiment was first raised and does not, as is sometimes said, commemorate the death of Sir John Moore. The shako may have been an undress alternative to the feather bonnet, or it might in fact have been worn during the latter stages of the Peninsular War as being more suitable than a hummel Kilmarnock bonnet. An inspection report before Waterloo noted that when on duty the officers were wearing their coats buttoned over, with shoulder plaids (as **G2**), and had blue web pantaloons and half boots.

H3: Sergeant-Major, full dress

Based upon an anonymous full length portrait dated to 1816, the sergeant-major wears an officer's style uniform, but is distinguished as the regiment's senior NCO by the four chevrons on his right sleeve and by the crimson worsted sash with a central yellow stripe.

Grenadier corporal and bandsman of the 79th, the latter wearing a white coat faced green. White coats seem to have been usual for bandsmen in highland regiments throughout the period covered by this book. (SUSM)

Notes sur les planches en couleur

A1 Le manteau habillé d'un revers vert et une broderie dorée est conforme aux régulations de 1768. Son plaid à ceinturon est en tissu écossais à dessin habituel gouvernemental ou de Black Watch. L'addition d'une rayure rouge au tissu utilisé pour le kilt est un détail retenu de l'uniforme des régiments parents. **A2** L'officier porte son uniforme de grande tenue avec une glaive, et sa garde très travaillée, fabriquée à Sterling mais n'ayant pas de dessin officiel. Son plaid à ceinturon, figurant probablement des plis cousus à l'avance, est d'un tissu simple gouvernemental. Le soldat du bataillon de la compagnie porte un petit kilt en tissus gouvernemental à rayure rouge. Les bas sont tricotés, on peut identifier ceci par le haut retourné, et ne portent donc pas la bordure noire sur l'écossais qu'on associe d'habitude avec le régiment.

B1 Ce sergent est habillé en grande tenue des highland, il porte un plaid à ceinturon, aux rayures caractéristiques couleur chamois et rouge. La rosette portée sur le béret est en broderie régimentale. Il porte aussi une écharpe d'officier. **B2** Les envelopes en chamois sont blanchâtres et on choisit un tissu écossais à rayure blanche, tissé en soie. Le sac à dos est en toile. **B3** Le tissu écossais unique du 79ème introduit une rayure jaune à un dessin écossais qu'on sait être des environs.

C1 On peut distinguer les 1ère Fencibles à partir de la rayure rouge dans leur tissu écossais. Les Membres de la Compagnie Légère ont les galons brodés, comme il en est coutume, et portent des bérets de Kilmarnock bleu uni. **C2** L'uniforme ne sert pas seulement comme petite tenue, mais est aussi portée par les régiments des highland en service aux Indes. Une garde est portée sur la glaive, mais on préfère l'épée courte, moins lourde et plus pratique. **C3** Le béret est peu commun car les carreaux verts aux croisements des carreaux rouges sont remplacés par des carreaux jaunes. La rayure jaune dans les tissus écossais est sans doute dérivée des revers. On porte des gilets rouges au lieu des gilets blancs utilisés d'habitude.

D1 Il est possible que le 97ème ait porté le tissu au dessin gouvernemental habituel. On porte des culottes ou pantalons en tissu écossais pour la plupart des tenues. Il semblerait que le béret bleu uni de Kilmarnock ait été porté par toutes les compagnies. **D2** Cet officier supérieur porte une culotte de cheval, des bottes, une épée courte aux revers jaune citron qu'on associe aux régiments d'Aberdeenshire. **D3** Tissu écossais de dessin gouvernemental, comporte des carreaux jaunes encore aujourd'hui. Mais à l'époque les carreaux étaient tissés en plus petit et les couleurs étaient plus clairs.

E1 Les officiers des Breadalbane Fencibles et ceux de la 116è portent tous les deux des rayures caractéristiques sur le kilt. Le manteau de l'officier des Fencible semble avoir un revers blanc et une doublure noire. La garde sur la glaive est aussi caractéristique. **E2** Ce dessin en dès sur le béret nommé Sutherland Dicing, comporte des carreaux blancs au lieu de verts au croisement des carreaux rouges. **E3** On voit parfois des revers noirs maise le dessin en dès gris/noir sur les bas semble unique à ce régiment.

F1 On porte le revers bleu clair caractéristique de préférence au tissue gouvernemental. La broderie argentée indique qu'il est 'pipe major'. **F2** La seule différence entre ce tissu écossais et celui du 78è est l'arrangement du kilt. Des rayures couleur chamois. **F3** Des bérets de Kilmarnock bleu uni pour la petite tenue. Une glaive à demi-garde à dessin régimentale identifie le cornemuseur en tant que membre de la Light Company. Tissu écossais à rayure rouge.

G1 Tissu à dessin gouvernemental, les bérets plats sont d'habitude réservés pour la petite tenue. **G2** L'écharpe est portée par dessus l'épaule plutôt qu'autour de la taille. On trouve des boutons tout au long du revers du manteau. **G3** Uniforme de petite tenue typique, le pantalon en tissu écossais remplacé par le kilt et un béret de Kilmarnock uni. La bayonette est nécessaire à chaque moment. La silhouette est colorée pour mieux représenter un membre de la 42ème au revers bleu et un tissus écossais verdâtre.

H Les uniformes différents portés par les membres du régiment pendant la campagne de 1815. **H1** Représentant les quatres régiments portant le kilt (42è, 78è, 79è et 92è) dans les Low Countries en 1815. Ils est défendu de porter la visière de cuir sur le béret. Le plastron carré est ovale d'habitude. **H2** On voit le gilet rouge clair de la Compagnie. On trouve un fil noir dans la broderie argentée. Le shako est une alterntive au béret à plumes. Lorsqu'il est de service, l'officer porte le manteau boutonné avec les plaids aux épaules, et des brodequins. **H3** On peut distinguer le grade supérieur de l'officier par son uniforme, les quatre chevrons à la manche droite, l'écharpe en worsted cramoisi, habillée au centre d'une raie jaune.

Farbtafeln

A1 Der Rock entspricht den Vorschriften von 1768, mit grünen Aufschlägen und Goldborten. Sein gegürteter Überwurf ist das übliche Regierungs- oder Black Watch-Sett. Die Zufügung eines roten Streifens im Sett des kleinen Kilts stammt von der Uniform des ursprünglichen Regiments. **A2** Der Offizier trägt die komplette Gala-Uniform und einen Pallasch mit einem verzierten Knauf, der keinem offiziellen Muster entspricht. Sein gegürteter Überwurf, wahrscheinlich mit abgenähten Falten, ist das einfarbige Regierungs-Sett. Der einfach Bataillonssoldat trägt einen kleinen Kilt mit rotem Streifen. Er trägt gestrickte Strümpfe, identifizierbar durch den umgeschlagenen Rand, und daher ohne die schwarze Einfassung der Karos, die das Regiment ansonsten zeigt.

B1 Dieser Sergeant trägt die komplette Hochländer-Uniform mit gegürtetem Überwurf mit lederfarbigen und roten Streifen. Die Rosette an seiner Kappe besteht aus Regimentslitzen. Er trägt auch eine Offiziersschärpe. **B2** Die lederfarbigen Hüllen waren weißlich, und der gewählte Tartan hat weiße, aus Seidenfasern gewebte Streifen. Der Knapsack ist im üblichen Segeltuch-Stil. **B3** Der einzigartige Tartan des 79. Regiments führte auf einem vorhandenen lokalen Sett einen gelben Streifen ein.

C1 Das 1. Landwehrregiment (Fencibles) scheint einen roten Streifen auf dem Tartan gehabt zu haben. Angehörige der Light Company hatten die üblichen Litzenschwingen und trugen einfarbig blaue Kilmarnock-Kappen. **C2** Die Uniform war nicht nur für den Dienst bestimmt, sondern wurde auch von Offizieren der Hochland-Regimenter in Westindien getragen. Ein Pallasch mit Korbknauf wurde manchmal getragen, doch wurds der leichtere, weniger umständliche Degen vorgezogen. **C3** Die Kappe ist insofern unüblich, daß die grünen Quadrate am Rande der roten Karos durch gelbe ersetzt sind. Der gelbe Streifen im Tartan kommt wahrscheinlich von den Aufschlägen her. Rote Westen im Gegenatz zu den üblichen wißen.

D1 Das 97. Regiment könnte das übliche Regierungs-Sett getragen haben. Tartan-Pantaloons oder Hosen wurden in den meisten offiziellen Uniformen getragen. Alle Kompanien scheinen die einfarbig-blaue Kilmarnock-Kappe getragen zu haben. **D2** Dieser Offizier im Feld trägt Reit-Breeches, Stiefel und Degen sowie die gelben Aufschläge des Aberdeenshire-Regiments. **D3** Tartan mit Regierungs-Sett mit gelbem Streifen, wie sie auch heute getragen werden, obwohl die Karos damals wesentlich kleiner und die Farben heller waren.

E1 Offiziere der Breadalbane Fencibles und die des 116. Regiments trugen Kilts mit auffallenden gelben Doppelstreifen. Der Waffenrock des Fencible-Offiziers scheint weiße Aufschläge und eine schwarze Linie in den Silberlitzen zu haben. Ebenfalls auffallend der Schwertknauf. **E2** Dieses Würfelmuster auf der Kappe zeigt weiße anstatt grüne Quadrate neben den roten Karos; dies ist als 'Sutherland'-Muster bekannt. **E3** Schwarze Aufschläge waren bekannt, aber dieses grauschwarze Würfelmuster auf den Strümpfen dürfte es nur bei diesem Regiment gegeben haben.

F1 Auffallende hellblaue Aufschläge anstatt des Regierung-Setts. Die Silberlitzen machen ihn als Tambourmajor erkenntlich. **F2** Der einzige Unterschied in diesem Sett zu dem des 78. Regiments liegt in der andersartigen Kilt-Anordnung. Weiße und lederfarbene Streifen. **F3** Einfarbig-blaue Kilmarnock-Kappen für die Dienstuniform. Säbel mit Halbkorb nach Regimentsart zeigt, daß dieser Dudelsackspieler zur Light Company gehört. Tartan mit rotem Streifen.

G1 Tartan in Regierungs-Sett, flache Kappen, normalerweise nur für Dienstuniform. **G2** Schärpe über der Schulter getragen anstatt um die Taille. Waffenrock voll geknöpft. **G3** Typische Dienstuniform, Tartanhosen anstatt Kilt, und einfache Kilmarnock-Kappe. Bajonett stets vorhanden. Farben der Figur lassen ein Mitglied des 42. Regiments erkennen, mit blauen Aufschlägen und einem grünlichen Tartan.

H Im Feldzug von 1815 trugen Mitglieder des Regiments verschiedene Uniformen. **H1** Vertreter aller vier kilttragenden Regimenter (42., 78., 79. und 92.) 1815 in den Niederlanden. Sie durften keinen Lederschirm auf ihren Kappen tragen. Quadratische Brustplatte, normalerweise oval. **H2** Weste der Red Light Company. Schwarze Linie in Silberlitze. Tschako als Alternative zur Federkappe. Im Dienst waren die Waffenröcke der Offiziere zugeknöpft, und sie trugen blaue Pluderhosen und Halbstiefel. **H3** Offiziersartige Uniform; er ist kenntlich als der führende Unteroffizier des Regiments durch die vier Winkel am rechten Ärmel sowie durch die karmesinrote Kammgarnschärpe mit zentralem gelben Streifen.